Classic Marques

Volkswagen Transporter
and Camper 1949-1967

Richard Copping

NOSTALGIA ROAD

First published by
Crécy Publishing Limited 2014

© Text: Richard Copping 2014
© Photographs: All images copyright
Volkswagen Alktiengesellschaft other than
Camper brochures: Canterbury Pitt, Devon,
Moortown Motors, Westfalia

A CIP record for this book is available from
the British Library

ISBN 9781908 347268

Printed in Malta by Melita Press

Nostalgia Road is an imprint of
Crécy Publishing Limited
1a Ringway Trading Estate
Shadowmoss Road
Manchester M22 5LH

www.crecy.co.uk

Front Cover: An early publicity shot of the Micro Bus
De Luxe.

Rear Cover Top: The versatile Pick-up introduced in
1952.

Rear Cover Bottom: 1949 prototype Delivery Van
outside the still bomb damaged VW factory at
Wolfsburg.

CONTENTS

Some markets developed their own special models. This illustration, from the cover of an Australian market brochure, depicts, amongst others, the Container Van (far right). Rather than perch a steel box on a Pick-up base, Volkswagen of Australia built its version direct onto the chassis platform of the Delivery Van. The cost of the vehicle was disproportionately high, with the result that only 123 were built between 1962 and the vehicle's demise in 1968.

Introduction

With a production span in Germany of over 17 years and in excess of 1.8 million vehicles sold, nobody could deny that the first-generation Transporter, known colloquially as the Bulli or Splitty, was not a resounding success for Volkswagen. Its replacement in the summer of 1967 with a second-generation model heralded even greater heights and paved the way for a lineage that exists to this day.

Similarly, its regeneration from a basic van with windows into a Camper, the pioneering work of a number of independent firms both in Germany and elsewhere during the 1950s, virtually guaranteed its survival long past the days when it should have been little more than a sprinkling of rust in the corner of an ancient scrap-yard.

Ever so gently, year-by-year, the first generation Transporter has eclipsed the mighty VW Beetle as a sought after icon of a bygone age until, by the dawn of the new millennium, it was increasingly sought after with resultant prices rocketing to unbelievable sums for even the most dilapidated of survivors. Camper mania, something which includes all members of the Transporter family of models, shows no sign of waning.

Less than a year after the last first generation Transporter had rolled off the van factory assembly line in Hanover, Volkswagen's Director General of 20-years' standing died. Heinz Nordhoff's final 18 months in office had been marred by controversy born out of Germany's first post-war recession. Politicians anxious to save their own skins had hit out at Volkswagen for Nordhoff's determination to perpetuate the life of the Beetle, by then an elderly design of over 30-years' standing. An open sore developed between Nordhoff and the government as the Director General criticised policy designed to ease hardship as totally opposite to what was needed in the motor industry. Due to the state's part ownership of Volkswagen, signs of Nordhoff's failing health and his age (he was already past the customary retirement age), steps were taken to replace him with someone more likely to be responsive to the government's line.

Nordhoff's successor was in place when the Director General died in April 1968. Kurt Lotz's four-year reign at the top proved disastrous, ending in the hierarchy's refusal to renew his contract. Lotz, who lacked a motoring industry background, appeared to be possessed by the single aim of denigrating the Beetle and by implication its mentor. As profits tumbled and his hoped for Beetle successor failed to emerge, it was essential that history be rewritten, playing down Nordhoff's triumphs and illustrating the folly of his unprecedented reliance on one model for so many years.

The text on the cover of this 1952 brochure asks 'who drives the VW Transporter?' Fifteen years later, at the end of first-generation Transporter production in Germany, Volkswagen could reply to its own question with the answer of a resounding 1.8 million owners.

It was during this process that what is still regarded by many as the authentic history of the first-generation Transporter was born. Nordhoff's star was duly eclipsed and the fable of a hastily scribbled sketch, magically transformed into the first-generation Transporter, became sacrosanct fact.

It is the aim here to set the record straight, making chapter one by far the most significant, while also delving into the idiosyncrasies of VW Camper folklore that delight the devoted army of enthusiasts, while hopefully galvanising the attention of the more casual reader.

Birth of the Transporter

INTRODUCTION

Volkswagen's Director General, Heinz Nordhoff, was not renowned for the brevity of his addresses to workforce, press and public alike. What some might suggest was his wordiness in the instance of the birth of the Transporter, or particularly so the speech he gave on its press launch on 12 November 1949, has since proved fortuitous. The level of detail — recorded word-by-word for posterity — sets the record straight forever. The Transporter as we came to know it did not originate with a hastily scribbled sketch, the penmanship of ebullient Dutch entrepreneur Ben Pon, in the days when the British were looking after Wolfsburg. Rather, the Transporter was born in the back of a car during what would no doubt have been a tense meeting between Nordhoff and his recently appointed design engineer, Alfred Haesner. Nordhoff, one time director in charge of the Opel truck plant in Brandenburg, the largest factory of its kind in Europe, easily capable of turning out over 4,000 vehicles per month despite the deprivations of war, wanted a second string to his Beetle's bow. That vehicle was to be the Transporter, one shell that could be offered in a multitude of guises, designed to fulfil a wide range of roles beyond the capabilities of the Sedan.

NO ROLE FOR PORSCHE

Both Porsche and the Nazis, joint creators of the Beetle, played no part in the origins of the Transporter. Although Porsche had developed military versions of what Hitler had christened the KdF-Wagen (strength-through-joy car) in the form of the Jeep-like Kübelwagen and the ingenious amphibious Schwimmwagen, the closest he came to designing a goods-carrying version of the Beetle was an ungainly shed-like structure perched over the area where the back seats would normally have been. Defeat in war, meant that the Nazi-owned factory built specifically to produce the Beetle passed into the control, if not the ownership, of the British, while Porsche, his son Ferry and his factory manager son-in-law, Anton Piëch, all fled to their native Austria.

THE BRITISH, IVAN HIRST AND BEN PON

That defeated Germany was divided into four zones proved fortuitous for the factory that was quickly renamed Wolfsburg, as it lay in the area under the control of the British. The military government duly despatched a Yorkshireman, Major Ivan Hirst, to the badly bombed plant, with no specific orders other than to take control. The fascinating story of how Beetle production began once more has little relevance here, other than to note the gifted amateur's ingenious solutions ensured survival and in one instance sparked an idea that was to provide the origin of the Transporter legend.

Major Ivan Hirst, the British officer sent to Wolfsburg in the summer of 1945 with no specific orders. Innovatory, he did a great deal to rescue the factory from a most uncertain fate. However, his expertise was not in the same league as that of former Opel Director, Heinz Nordhoff.

Dutchman Ben Pon (left) with his somewhat less flamboyant brother, Wijand. His sketch of 1947 vintage is cited by many as the ancestor of the models launched in November 1949.

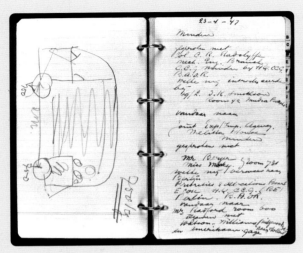

Ben Pon's sketch of April 1947 aroused Ivan Hirst's interest; he suggested it was presented to his superior, Charles Radclyffe. The Colonel's answer to the duo's suggestion that the Delivery Van idea should be progressed alongside the Beetle was a resounding no.

Deprived of transport to move materials from one part of the factory to another, Ivan Hirst created what became known as the Plattenwagen (literally 'flat car'). Taking the chassis of a Kübelwagen, or Beetle, he put a flat board at one end and a driver's seat over the engine at the other. The Dutch entrepreneur, Ben Pon, whose aim for many a year had been to sell Volkswagens in his country, spotted a Plattenwagen on one of his far from infrequent visits to Wolfsburg, and immediately saw a market for these vehicles as an upgrade to the three-wheeled delivery vehicles populating Holland. However, his request to the Dutch Transport Authority for a street legal certificate was immediately turned down, regulations stating that the driver must be seated at the front being the cause.

Temporarily down, but far from out, Pon bounced back with his now legendary sketch. The date was 23 April 1947, some seven months before Nordhoff's appointment and a time when the person with ultimate responsibility for Wolfsburg, was Colonel Charles Radclyffe who was based at CCG (Control Commission for Germany) located in Minden and who held responsibility for all engineering construction in the British zone.

The essentials of the sketch were a box-shaped vehicle of 750kg, with an apparently contoured front and its driver and any passengers sitting over the front wheels. The engine, assumed to be Porsche's air-cooled flat four and borrowed from the Beetle, was mounted above the rear axle and appeared to be easily accessible via a large upward opening lid.

Ivan Hirst was captivated by what he saw and immediately recommended Pon met Radclyffe. However, this was not to be, as the colonel was only too aware of the ongoing uphill struggle to maintain even basic Beetle production amidst a factory still lacking the basic essentials. He was clearly of the opinion that a second Volkswagen at such a stage was one step too far.

NORDHOFF, HAESNER AND THE ROAD TO TRANSPORTER PRODUCTION

Appointed earlier in 1948 as head of technical development, Alfred Haesner was given less than a year to produce a new vehicle for Volkswagen. Well-equipped for the task, having previously designed commercial vehicles for Phänomen in Zittau, it is abundantly clear that he knew exactly what Nordhoff wanted from remarks he made as the prototypes began to take shape:

'Designed for town and country, short and long distances, autobahns and field tracks, goods and passengers, retail and industry ... This commercial vehicle is suitable for all sectors of business, express delivery and freight transport, for example a minibus, special purpose vehicle, post van, ambulance or mobile shop.'

Heinz Nordhoff: Volkswagen's Director General 1948-68 and the Transporter's mentor. This photograph was taken in the mid-1950s, one of a series designed to accentuate Nordhoff's profile at home and abroad.

Heinz Nordhoff's strategy was quite simply one of making Volkswagen the shining example of Germany's post-war recovery and as quickly as possible. Apart from putting tremendous pressure on his staff, his enormous capacity for gruelling schedules of work ensured that what he described as a 'desolate heap of rubble' was rapidly transformed into a highly efficient, modern factory. Similarly, 'the amorphous mass which lacked discipline or principle, had no factory organisation in a real sense, and was without a programme or rational work system' was changed beyond recognition. The Beetle was shaken and stirred into a desirable export-winning car, one that would generate valuable income for Nordhoff's Volkswagen. The De Luxe was launched in July 1949; Nordhoff wanted a second income-generating vehicle to be ready in the same year. Pon's delivery van would not have been sufficient.

Below and below right: Prototype Delivery Vans as photographed and included in the very early Transporter brochure entitled *VW Vans — a Picture Library*. Enthusiasts love to pore over details such as the single wiper, unusual headlamp surrounds and vertical air intakes (at the rear of the vehicle), all features that were modified before series production began.

Having been tasked with creating the Transporter (project EA-7) in the autumn, by 11 November 1948 Haesner felt compelled to request more staff. Nevertheless, within nine days, he could present Nordhoff with two design outlines. Both bore similarities to Pon's by then overlooked sketch, but each could be justifiably described as the result of extensive market research. Version A had a flat front with a roof overhang, while Version B, which Nordhoff instantly preferred, had a slightly curved and raked cab. By 9 March 1949, 1:10 wooden scale models fitted with three interchangeable nose shapes had been subjected to wind tunnel tests. The flattest of the cones indicated a CW of 0.75 and 0.77, while the rounded version offered 0.43.

The first full-size prototype of what was known in the factory as VW29 was ready for testing on 11 March 1949. Aware that money was tight, Haesner had been persuaded to believe that the Beetle's chassis, appropriately widened to carry the box-like structure of the prototype, would be sturdy enough to carry loads heavier by 50% and more, simply because as with the Sedan, the body and framework were bolted to each other.

Tested at night, in an attempt to maintain a degree of secrecy, Ivan Hirst, who was still at Wolfsburg, but now acting as custodian before an official handover to the German government due to occur later in the year, remembered many years later that when the vehicle came back on 5 April: 'it was six inches lower. The weight of the load broke the back of the flat-section at the centre of the platform frame.'

Perhaps not surprisingly, Nordhoff would tolerate no delays. On 19 May, he set a start date for production of 1 November or, as an absolute latest, 1 December. His immovable intention was that the vehicle should be ready for an official launch at the Geneva Motor Show to be held in March 1950. By that time, the variants on hand must not only have been both thoroughly tested at Wolfsburg, but also pre-production models made available for appraisal by Volkswagen's most valued customers.

Amazingly and despite Nordhoff's insistence on being involved in every stage of the development process, a second prototype was available on the very day the Director General declared his time-scale for launch. Both this vehicle and the rebuilt first prototype now featured a bespoke frame of a style that in future years would become known as unitary construction. Torsional rigidity was substantially improved, for the floor and body were now firmly welded together. Five substantial cross-members were welded between the front and rear axles, while two hefty longitudinal rails and a series of outriggers guaranteed additional strength to the whole structure. Rigidity beyond need was achieved by welding the steel floors in the cab and cargo area to the platform, as well as the one above the engine compartment. Additional work included strengthening both the front and rear shock absorbers.

As financial constraints precluded the development of a more powerful engine than the 25PS unit offered with the Beetle, Wolfsburg's archive provided at least a partial solution to the Transporter's distinct lack of acceleration. Employing the Kübelwagen's 1:1.4 reduction-gear rear hubs proved satisfactory, while an adjustment in the height of the front suspension also created the benefit of increased ground clearance. Dual torsion spring units were employed for the back axle as a simple way of keeping costs down.

When the prototypes performed well within the confines of Wolfsburg, Nordhoff decreed that they should be rigorously tested on some of the worst roads in rural Saxony. Having duly accomplished some 12,000km without serious mishap, Nordhoff added further pressure by demanding that a wider variety of vehicles should be available by 15 October. These were to include a pick-up, an eight-seat minibus and an ambulance, plus a vehicle for the Bundespost to use.

As selected prototypes began to emerge, each illustrating improvements over earlier examples, Nordhoff continued to oversee what appeared to be the minutest details. He stipulated further improvements, changes such as stronger door hinges, lighter-weight doors, and the lowering of the engine bay 'roof' panel to allow for an extra layer of insulation, which would reduce the amount of heat transmission from

Two publicity shots of the prototype Delivery Van, both of which were taken within the grounds of the Wolfsburg factory (note how many of the building's windows had yet to be re-glazed). The external filler cap, if nothing else, confirms the vehicle's prototype status.

the engine to the interior. Seeming trivialities included an order to fit a second wiper so that the screen in front of a passenger could be kept rain free, and a direction to improve both heating and ventilation. Due to the danger that petrol might be siphoned off (a practice rife in post war Germany), Nordhoff instructed that the external cap must be relocated to a position within the engine compartment, just as the Beetle's was hidden in the car's boot.

Further wind tunnel work took place in September 1949, again using models. Although the Transporter was no longer as streamlined as it had been, the resultant CW of 0.44 was deemed acceptable, much to the relief of an agitated Haesner who knew that if the results had been unfavourable, Nordhoff would have demanded further work. Likewise, as late as 26 September, last minute working around the clock was not to be avoided when appropriate roof pressings could not be located. With a bevy of authorised overtime sanctioned, the press plant managed to achieve what was required, but even then, painting took longer than envisaged, the roofs finally being ready on 20 October.

A final week's delay put the press launch back from 6 to 12 November, as Nordhoff continued to make further changes. On the day, Nordhoff was high in his praise for Haesner and careful, despite Pon's presence in the audience, to reiterate that this totally new concept was his work:

'With our new Transporter, we have created a vehicle the like of which has never been offered in Germany before. A vehicle which had only one aim: highest economy and maximum utility value. A vehicle that did not have its origin in the head of engineers but rather in the potential profits the end-users will be able to make out of it. A vehicle that we don't just build to fill our capacity — that we can achieve for a long time with the Volkswagen Sedan — but in order to give the working economy a new and unique means to raise performance and profit.'

The prototype Micro Bus — proof positive, if such was required, that from the start Nordhoff wanted a range of vehicles rather than just a Delivery Van. This vehicle was present at the Transporter's press launch in November 1949.

Range Building
— FROM THE DELIVERY VAN TO THE MICRO BUS DE LUXE AND PICK-UP

INTRODUCTION

True to Nordhoff's intention, when the all-new Transporter made its appearance at the Geneva Motor Show in March 1950 not one model, but three were present to meet the admiring glances of an attentive audience. Apart from the Lieferwagen or Delivery Van, there was the highly significant Kombi, a dual role model as its name suggests, and top of the initial range, the VW Achtsitzer, or eight-seat Micro Bus as it would be known in many parts of the world in later years. More quickly followed, both in the guise of core range models and as vehicles specially adapted for specific uses. Such was their extensive nature that from the humble beginnings of Sonder Packung (special sets) for the Transporter, which were listed as early as 1951, by 1956/57 SO codes had been devised. SO stood for Sonderausführungen, or special model, a device used for a combination of specific purpose vehicles built at the factory, or in many more instances, adapted by approved specialist karosseries (coachworks). In one or two cases, a vehicle that had started as a special model was drawn into the fold of a later date expanded core range, Volkswagen dismissing the karosserie involved to produce, for example, the Double Cab Pick-up at the factory.

THE DELIVERY VAN

Nordhoff's statement at the press launch that research had shown that customers did not want the market norm of a 'typical half-tonner on a car chassis ... but a 50% bigger three-quarter-tonner van which can be used in many different ways' summarised both the originality and ingenuity of the Delivery Van's design. Vans based on a car's chassis, he asserted, inevitably suffer as the 'load area lies completely over the rear axle, which carries practically the entire payload and therefore soon reaches its natural limit'. Volkswagen's Transporter was entirely different. 'We started from the load area,' declared Nordhoff, which was 'actually much more obvious, and original. This load area carries the driver's seat at the front, and at the rear both the engine and gearbox — that is the patent idea, free of compromise, for our van.'

Using early publicity material as an official guide, it is easily confirmed that the Transporter measured no more than an additional 50mm in length when compared to the Beetle, the former totalling 4,100mm and the latter 4,050mm. However, as

As intended — an array of Transporters of different varieties on the Volkswagen stand at one of the many motor shows held in 1952.

The cover of one of the earliest of brochures produced to promote the Transporter. Amazingly, although not obvious here, the unknown artist based his sketches on prototype models.

Early production model Delivery Van — note the lack of both a rear window and bumper. The massive VW roundel was a feature of all Transporters until 11 November 1950.

expected, the Transporter was both wider and taller to the extent of 120mm and 400mm respectively, the difference in height (1,500mm versus 1,900mm) between the two vehicles doing most to feed the illusion that the sedan and van were in different leagues when it came to size.

As for the Delivery Van's generous load-carrying capabilities, Nordhoff ensured this was not overlooked when he cajoled and regaled the press in November 1949:

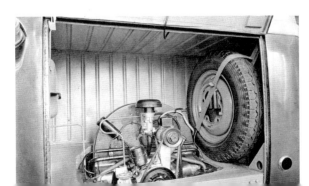

The 25PS engine dwarfed by the size of the engine compartment — the spare wheel was positioned horizontally on a platform above the engine from 1951.

'The Transporter comprises a main area of three square metres of floor space plus, over the engine, an additional square metre and 45 cubic metres of volume. In short, neither the load area nor the driver's area is restricted by these items.'

As presented until March 1955, the Delivery Van and most other members of the range lacked any form of rearward access, the design centring on an exceptionally large engine lid, which gave rise to such models being given the nickname of 'barn door' in later years. Brochure copywriters managed to turn what might have been considered a design defect linked to an engine at the rear of the vehicle to their advantage by suggesting a customer preference for side-loading doors. Admittedly, access from the pavement both for loading and unloading was convenient, particularly when the parking space available was restricted and would have prevented opening the rearward doors properly. For greater flexibility, the option to specify a second set of side-loading doors on the driver's side of the Delivery Van, while at the same time fitting underbelly floor plates, came in June 1951. However, it would be April 1963, before the Delivery Van could be specified with a sliding side-loading door.

An early (but post April 1951) Delivery Van at work in this carefully posed shot. The size of the engine lid is obvious. The presence of a rear window helps to date the photograph.

Once the American advertising agency Doyle Dane Bernbach (DDB) became involved in the production of Volkswagen's publicity material and brochures, a truly radical style evolved that other manufacturers were anxious to copy. Here's the agency's typical no-nonsense approach to the optional second set of side-loading doors.

Relevant to all models is the Transporter's distinctive look; a smiling personality created through the 'Y'-shaped swage lines at its front and above all by the split pane windscreen. While the former amounted to no more than a design detail that may or may not have been attempted to give a slight family resemblance to the Beetle, the latter was undoubtedly a matter of expedience. Wind tunnel testing had shown that a more curved front was judicious, but at the time to produce a curved windscreen was relatively difficult and undoubtedly costly. Mimicking the Beetle's split rear window, produced in such a format for exactly the same reasons, appeared an appropriate solution. While by the 1960s, the Transporter began to look distinctly old-fashioned, the nature of the divided panes only adds to the vehicle's appeal today.

THE KOMBI

The Kombi made its production debut on 16 May 1950 and, as its name implies, could act as a delivery vehicle one minute and as a means of transporting a family of up to eight the next. Better still, with fewer people to house, it could carry out both tasks at once.

Nordhoff reasoned that there would be a large number of people who, unable to afford both a car and a Delivery Van, would be delighted with such a dual-purpose vehicle and he was proved correct virtually from day one.

The Kombi's key selling point was, as its name implies, its ability to be a Delivery Van one moment and a practical people carrier the next. Artist Bernd Reuters was on hand to capture such a message in a most successful manner.

Although a 1960s approach to promoting the Kombi undoubtedly catches the eye, there is no getting away from the basic nature of the vehicle. Note the side-loading doors and their lack of trim, while the glimpse afforded of the interior again reveals mean austerity.

Although popular, the Kombi was far from luxurious. As with the Delivery Van, bare painted metal was in abundance. Essentially, the only changes to what might still be described as the load area, were the three rectangular windows cut into either side of the vehicle and rubbing matting, of the style fitted in the cab, to cover the floor. Even the utilitarian type of hardboard headlining fitted in the cab was missing.

As for the two bench seats, these were basic both in the way they were fitted and removed, the provision of wing nuts being deemed sufficient, and in their nature, with sections of the frames being exposed for all to see.

The importance of the Kombi and more particularly its versatility should never be underrated, for the concept spawned a further incarnation of the Transporter, possibly the most sought after today.

THE VW ACHTSITZER

Following testing by a valued customer, which had begun on 16 April 1950, the Achtsitzer went into full production just nine days after the Kombi, on 22 May. Also known as Der Kleinbus, literally the small bus, in Britain it would become the Micro Bus and in the USA, the Station Wagon. In February 1956, the addition of a seven-seat version, the VW-Siebensitzer, did little to clarify which title was preferable!

Compared, to the Kombi, here was sheer luxury, opulence that started with a soft cloth headlining (which extended to the window surrounds) and then worked its way across all aspects of the vehicle. Fibreboard panels clad with vinyl adorned the interior below the waistline (including the doors), while pleated and piped seat upholstery, plus partial vanity boards and a sliding partition over the engine compartment to conceal any luggage stored at the rear, all contributed to the ambience of luxury. Even the rubber floor covering differed, being of a raised square design matting.

Four early images of the startling new Micro Bus. The photographer took care to ensure the message of the vehicle's seating and passenger-carrying abilities were at the forefront, although in one instance the association between luxury travel and the Transporter was highlighted.

Opposite: A 1960s approach to the Micro Bus places emphasis on the number of people who can travel comfortably in such a vehicle, while also illustrating that copious amounts of luggage can be securely loaded through the vehicle's rear hatch.

Externally two-tone paintwork, in preference to a single colour, could be specified, although even the standard single shade varied from those available on other models. By way of a side note on the subject of paint, Volkswagen gave customers the choice on all models of a vehicle finished in nothing more than primer. Ludicrous as this might first sound, the option proved popular, with 2,398 of a total production number of 6,462 vehicles being despatched from Wolfsburg in this state in the first year of production. Delivery Vans could be painted in a firm's house colours and logoed accordingly and, to a lesser extent, this was the case with both the Kombi and Microbus. Such was the demand that as early as 1951 Volkswagen published a brochure entirely devoted to painted and sign-written vehicles that would have left the factory in their underclothes.

Bernd Reuters appears to have been tasked with preparing both simple portrait and, on occasion, highly sophisticated artwork. This is the delightful cover of his early 1951 brochure promoting the Micro Bus. Sadly, it was a short-lived affair, as another Transporter of a more fitting nature to warrant special artwork was soon to evolve.

Scene setting! Wouldn't *you* like to be embarking on a skiing holiday in your Volkswagen Micro Bus?

Amazingly, photographs of the very first vehicle to be finished in primer (and nothing more when it left the factory) survive. Despatched on 8 March 1950 to Autohaus Fleischhauer in Cologne, the Delivery Van was soon adorned with the livery of the 4711 Perfume Co. Note also the chromed bumpers and lack of the VW roundel on the front of the vehicle.

Above: Vehicles finished in primer proved very popular and Volkswagen quickly issued a brochure packed with examples of vehicles adorned with company liveries.

A 1960s take on livery or special paint jobs.

Der **VW** *Kleinbus*

Reuters' portrayal of the Micro Bus De Luxe for the cover of an elaborate, multiple page brochure, covering both the latest model and the less expensive Micro Bus, probably carries the status of being the artist's best Transporter work for Volkswagen. Note particularly how he reduces the size of the occupants in relation to that of the vehicle, the streamlined appearance of its panels and even the speed lines by its wheels.

VW KLEINBUS — 'SONDERAUSFÜHRUNG'

Over a year passed by before a further variation on the Transporter theme emerged and with good reason, for the next model carried a hefty price tag appropriate to its luxurious nature. While Nordhoff had been determined to offer a De Luxe version of the Beetle in the summer of 1949, his motive for delaying the equivalent in Transporter form was perfectly understandable. The De Luxe Beetle provided a car worthy of export and in turn, much needed revenue. The basic or Standard Beetle had little appeal outside Germany. However, a De Luxe version of the Transporter would never be a mainstay of Volkswagen's success overseas. Nordhoff, therefore, bided his time until the Germany economy had revived and at least a proportion of its habitants had sufficient disposable income to contemplate a top-of-the-range model.

Publicity shots of the Micro Bus De Luxe are, as might be anticipated, plentiful in number. These two are amongst the earliest and portray the vehicle with a white roof panel, a feature that was not part of the specification when production started in earnest.

Although still taken in the 1950s, these three images date to the period after all Transporters were revamped, an event which occurred in March 1955. The message pertaining to luxury travel is still just as pertinent. Note how the vehicle at the airport (a favourite setting) has been specified without roof skylights.

The Micro Bus De Luxe, or 'Special Model', first rolled off the assembly line on 1 June 1951. Externally, it could be distinguished by a near full length canvas, fold-back sunroof and a series of eight skylight windows in the roof panel, each glazed with the Perspex type material known as Plexiglas. In addition to four virtually square windows (rather than the Kombi and Achtsitzer's three oblong ones), Plexiglas was used again to create wraparound windows on each rear quarter. These met with a much larger rear window than was the norm (any form of rear window only having been added to the other models' specification in April) in the panel above the barn door engine lid. With 23 windows in total, plus the aforementioned sunroof (which became an extra cost option for Micro Bus purchasers at the same time), the De Luxe was both light and airy.

Both externally and internally, the trim package was much enhanced. Apart from a tentative attempt to provide the vehicle with a form of rear bumper (it would be December 1953 before the other models had any form of protection from parking knocks) chunky bright-work trim strips adorned the waistline and 'Y'-shaped swage line on its front, while the large

Regular size.

Large economy size.

The skill of the DDB copywriter in the 1960s could not be challenged. This advert, one of a series collected and presented in brochure format, outlines the similarities between the Beetle and the Transporter. It states very little which is specific to the model. The theme in reality is one of the number of seats required. Here's the copywriter's concluding statement: 'Once upon a time people had trouble deciding whether to buy a VW or not. Now they have trouble deciding which size.'

A further 1960s carefully posed 'action' shot. Apart from telling the beholder that the Micro Bus De Luxe could hold many passengers, the image gives a glimpse of the comparative luxury associated with the top-of-the-range vehicle.

How many school buses would really be the pricey Micro Bus De Luxe model?

Count the people catching a moment to take in the beauty of the view (only available via a VW Micro Bus De Luxe, of course).

VW roundel was chromed, as were the hubcaps. Although the De Luxe could be specified in the shared Micro Bus shade of Stone Grey, it was also offered in the most delightful of two-tone paint combinations, namely Chestnut Brown over Sealing Wax Red. Inside, the De Luxe sported a full-length dash, all other models being reliant on a single binnacle. The dashboard carried the luxury of a large mechanical clock, as well as offering ample space for the fitting of a radio. In the mould of the Export Beetle, fittings and fixtures were offered in an ivory colour, rather than the standard black of other models. Upholstery was both pleated and piped as might be expected, while the rear luggage compartment was not only carpeted but also featured both chromed rubbing strips and protective restraining bars over the lower parts of the surrounding windows.

Predictably, the Micro Bus De Luxe was not cheap and in 1957, when all the core members of the Transporter range had been established for a few years, only the special purpose Ambulance cost more. (Kombi DM 6,600, Micro Bus DM 6,975, Micro Bus De Luxe DM 8,475, Ambulance DM 9,500)

VW-KRANKENWAGEN

From the earliest days, Nordhoff had included an Ambulance in his list of what he perceived to be key models. In the event, the coachbuilder Miesen was quicker off the mark, although its Transporter based Ambulance lacked a key feature of the product launched by Volkswagen on 13 December 1951. As production figures will later reveal, and as no doubt anticipated, total yearly production was paltry in comparison to that of all other models. Thanks to its specialist nature, the Ambulance's inclusion in many full Transporter line brochures must have baffled many glancing through such material.

What was exceptional about Volkswagen's Ambulance was the amount of work and attendant cost required to make the vehicle practical for its role. While, as Miesen hoped would be the case, medical attendants could manage to manoeuvre a stretcher through side-loading doors, a vehicle with a rearward entry point would be much easier to use.

In order to achieve this, Volkswagen reduced the size of the engine compartment (and its attendant lid), relocating both the resident petrol tank and spare wheel in the process. The filler cap for the former became an external arrangement, while the spare wheel found a home behind the backrest of the cab bench seat. Initially, the new hatch opened upwards, providing shelter as a stretcher was manipulated through into the body of the Ambulance, though within a short time it was redesigned to drop down, thus offering a ramping arrangement to help ease patient and apparatus into position. It was at this point that of necessity the window in what had briefly been a hatch was deleted from the specification.

Reuters was ready with his easel and paints to portray the new Ambulance in a flattering light. As the vehicle's most outstanding feature was access from the rear for stretchers, the artist turned the vehicle to show this asset to full advantage. A night-time portrait also helped when depicting frosted glass.

Partially frosted side windows protected by rails, a step-up arrangement through the side loading doors and a host of specialist equipment, served to merit the Ambulance's premium price. As a bonus for Volkswagen some of the worked carried out would be useful when it came to the evolution of the final core model, the Pick-up.

VW-Pritschenwagen 'Pick-up'

Again much talked about from the earliest of days, the sales potential for the Pick-up should have demanded its launch earlier in the programme. Essentially its seemingly delayed arrival related to the costs involved in its initial production, namely the design and expensive retooling required to produce a flatbed vehicle. The Pick-up finally made its debut on 25 August 1952.

Fortunately, some of the work required had already been done in respect of the Ambulance; the engine compartment had been reduced in size, the fuel tank and spare wheel relocated. That left Volkswagen with the task of creating a flat bed, and designing completely drop-down sides to make the task of loading and unloading as easy as possible. The necessary relocation of the engine's cooling louvres was relatively simple, and while the creation of a lockable storage space

As with the Ambulance before it, Reuters turned the Pick-up so that the loading platform could be seen in all its glory. Note how the artist also illustrates the secure storage area between the engine and the cab, while it is clear that the hinged side flaps could be fully lowered, thus making loading or unloading particularly easy.

Some of the photographic images taken of even the most mundane of vehicles were truly stunning once the DDB agency was responsible for Volkswagen's publicity. Simple in concept, direct in message, but so, so effective.

where valuables could be concealed out of sight was ingenious, in design terms the task was not that difficult. The most expensive part of the process was the creation of a new pressing for the vehicle's truncated roof panel.

However, the expense and effort involved in the creation of the Pick-up quickly proved worthwhile. In its first full year of production, sales exceeded those of the Micro Bus and its De Luxe counterpart combined and were in touching distance of those of the Kombi. Furthermore, the Pick-up along with the Delivery Van proved indispensable as a vehicle that could be adapted for special uses.

SONDERAUSFÜHRUNGEN — SPECIAL MODELS

The subject of special models has already been touched upon in the introduction. All that remains necessary here is to emphasise the importance of such models to Volkswagen, to note that both model specific brochures and more generic publications,

A page from a 1952 brochure design to depict the wide variety of vehicles put to special uses. The images are themed, as can be detected from the bottles depicted to the right of the image. Here there is a Delivery Van dispensing a particular brand of soft drink and a Pick-up adapted to carry many more crates of drink than the standard flat-bed would permit.

VW-Verkaufswagen (SO 1)

Plain and simple: the mobile shop SO1 with a large hatch cut into the panels of the side of the vehicle that would normally be adjacent to the pavement.

SO13 — Pick-up with enclosed storage box.

covering a host of possible ideas for the Transporter, were produced and to list some of the finest examples. Additionally, it is worth noting that by 1961, Volkswagen could list 130 variations with official SO (special model) status, while many others not afforded such standing, were in regular production. In the early days, the SO designation also carried the name of the town where the coachbuilder was located and it is for this reason that the contents of a simple brochure, presented by the Munich dealer Mahag and printed in 1957, are repeated here:

SO1 Wiedenbrück	Mobile Shop
SO4 Wiedenbrück	Traffic Accident Command Vehicle
SO8 Marburg	Wide-bed Pick-up (metal)
SO9 Wiedenbrück	Wide-bed Pick-up (wood)
SO11 Hagen	Pick-up Ladder Truck
SO12 Minden	Pick-up Box Wagon with Aluminium Roll Shutter Doors
SO13 Wiedenbrück	Pick-up with Enclosed Storage Box
SO14 Winsen	Pole Carrying Trailer and Mounting for Load Bed
SO16 Lorch	Pick-up Double Cab conversion
SO18 Frankfurt	Mobile Workshop, Breakdown Vehicle
SO22/SO23 Wiedenbrück	Westfalia Camping Box and Westfalia De Luxe Camper

In the foreground, a standard 1960s Pick-up. In the background an M201 (M standing for mehrausstattung or optional equipment); the latter was a wide-bed Pick-up with wooden loading platform and side panels. This variation on the Pick-up's standard specification, rather than being a special model in its own right, was produced by Westfalia.

Popularity and Improvement

INTRODUCTION

This is one of those rather matter of fact chapters, although hidden in its depths are two additional models deemed worthy of inclusion in publications describing the core range.

The popularity of the Transporter can be seen in the ever-increasing numbers in which each model was built (with the possible, but realistic, exception of the Ambulance), at least until the mid 1960s. Similarly, wherever the Beetle went (or was exported) the Transporter tended to follow. Some countries achieved so many sales that they contemplated the assembly of CKD (Completely Knocked Down) kits, and in a few instances, Brazil, Australia and South Africa most notably, turned to full manufacture.

Such was the demand for both the Beetle and the Transporter, and in the instance of the former, an inability to keep pace with that requirement, no matter how much money was invested in increasing production, that it was deemed necessary for manufacture of the Delivery Van, Kombi et al to move elsewhere; to a purpose-built factory in Hanover.

Inevitably, a vehicle with a production span of 17 years would be modified during its lifetime. Nordhoff was renowned for breaking with tradition and not replacing his Beetle with a new model every few years. Instead, he believed in a policy of continual improvement, applicable to all types of Volkswagen produced in his lifetime.

It's 9 October 1954, Wolfsburg and the occasion of the 100,000th Transporter to be produced rolling off the assembly line. Nordhoff made the customary speech and, no doubt, already had it in his mind to move manufacture elsewhere.

Although sales in Germany were strong, it was exports that made the Transporter (like the Beetle) the phenomenon it quickly became.

PRODUCTION NUMBERS

The production numbers shown in the chart below represent vehicles built first at Wolfsburg and later at Hanover, in other words German production, but exclude vehicles manufactured at any of Volkswagen's satellite operations. Such Transporters totalled in the region of 430,000 vehicles, when production finally ceased in Brazil in 1975.

Year	Total	Delivery Van	Kombi	Micro Bus	Micro Bus De luxe	Pick-up	Ambulance
1949	8 prototypes	6	1	1			
1950	8,059	5,662	1,254	1,142		1	
1951	12,003	6,049	2,843	2,805	269	1	36
1952	21,665	9,353	5,031	4,052	1,142	1,606	481
1953	28,417	11,190	5,753	4,086	1,289	5,741	358
1954	40,199	14,550	8,868	5,693	1,937	8,562	589
1955	49,907	17,577	11,346	7,957	2,195	10,138	694
1956	62,500	22,657	16,010	9,726	2,072	11,449	586
1957	91,983	30,683	23,495	17,197	3,514	16,450	644
1958	101,873	36,672	21,732	19,499	4,342	19,142	486
1959	121,453	41,395	25,699	22,943	6,241	24,465	710
1960	139,919	47,498	30,425	22,504	7,846	30,988	658
1961	152,285*	45,121	35,950	25,410	8,095	36,822	883
1962	165,774**	47,237	38,506	29,898	11,280	38,118	728
1963	174,866	47,891	40,882	31,196	14,764	39,458	675
1964	187,947	48,481	44,659	40,115	14,031	39,832	829
1965	176,762	43,723	44,331	37,933	12,467	37,444	864
1966	176,275	43,084	46,284	30,767	18,790	36,534	816
1967	68,100***						

*Includes four others,

**Includes 7 others,

***Changeover year, August, model breakdown not possible

THE MOVE TO HANOVER

The decision to move Transporter production from Wolfsburg to Hanover was made by Nordhoff on 24 January 1955. Although a daily output of 170 models and total production for the previous year of just over 40,000 Transporters was trivial in comparison to that of the Beetle, with over 200,000 cars manufactured annually, a momentum for growth was evident. However, on 9 October 1954, the 100,000th Transporter had rolled off the assembly line and Nordhoff rightly predicted that it would only be a relatively short time before such a number was produced annually.

Just the kind of publicity picture Nordhoff approved of. Here he is (third from the right) personally overseeing building works — a frequent occurrence during his 20 years as head of Volkswagen.

The foundation stone laying ceremony for the new factory took place on 1 March 1955 and just over 12 months later the first vehicle, a Dove Blue Pick-up, was gently eased off the assembly line. Mass production at the 112 hectare Stöcken site on the outskirts of Hanover began officially on 20 April. The 200,000th Transporter followed on 13 September 1956 and, just six years later, on 2 October 1962 the millionth vehicle, a heavily garlanded Micro Bus De Luxe, took its place in history amidst much pomp and ceremony.

It's now 1956 and here is the new Hanover factory in all its glory — the western elevation and administration building extended to some 378m.

Right: On 9 March 1956 Otto Hoehne (left), head of the new Hanover factory, handed over the keys of the first Transporter, a Dove Blue Pick-up, to roll off the assembly line.

Left: It's 20 September 1962 and the heavily garlanded millionth Transporter stands by the side of a flower bedecked podium.

The millionth Transporter leaves the Hanover factory, one of the 750 vehicles produced daily, many of which were destined to populate in the region of 130 countries across the world.

YEAR-BY-YEAR IMPROVEMENT

A summary of the most significant changes made to the Transporter's specification.

Date	Change
1950, June	Partition added between cab and load area of the Delivery Van.
1951, April	Small rear window added to previously blank panel above engine lid.
1953, March	Synchromesh standard for all models on 2nd, 3rd and 4th gears. Full rear bumper fitted to Micro Bus De Luxe.
1953, December	Engine upgraded from 25PS to 30PS with an allowable top speed of 50mph. Rear bumper fitted to Delivery Van, Kombi and Micro Bus.
1954, April	Fuel gauge standard for the Ambulance, Pick-up fitted with rear bumper.
1955, March	All models given a major facelift. Redesigned roof panel with peak over windscreen to accommodate air intakes for a much improved ventilation system. Full width metal dashboard replaced single binnacle. Redesign of rear of vehicle reducing the size of the engine compartment and its lid substantially (in line with the Ambulance). New rear hatch (900mm x 730mm) afforded rearward access for all models for the first time. Spare wheel behind cab seat, wheels reduced in size to 15in from 16in. Overall load space in the Delivery van increased to 4.8m².
1958, May	Single central brake-light, replaced by new twin larger tail lamps incorporating brake-light function.
1958, August	Stronger bumpers fitted. US market fitted with two tier bumpers to align Transporter's with those of the average American vehicle.
1958, 3 November	Production of the Double Cab Pick-up started (replacing the Binz special model). Re-tooling involved the creation of a larger roof panel than that of the single cab, a side door allowing access to the second row of cab seats (or extra storage space if these were removed), plus the creation of a truncated loading platform, which involved losing the under bed storage area. A new hideaway was created though, in the form of concealed storage under the second row of cab seats.

The Volkswagen Transporters

Reuters accentuated the changes made to the Transporter's frontal appearance in March 1955 to near ridiculous proportions, in what was to prove his last new work relating to the Type 2. The peak at the front of the roof panel concealed vents designed to stop the severe misting-up problems experienced previously.

1959, May	30PS engine redesigned with higher compression ratio. Stronger crankcase halves, sturdier crankshaft. All synchromesh gearbox fitted.
1960, June	34PS engine fitted – essentially the same as the one introduced in 1959, but with increased power due to new carburettor 28PICT. European models antiquated semaphores replaced with flashing indicators (US models had benefitted from this arrangement since 1955).
1961, July	Fuel gauge finally standard on all models.
1961, August	Larger tail light clusters.

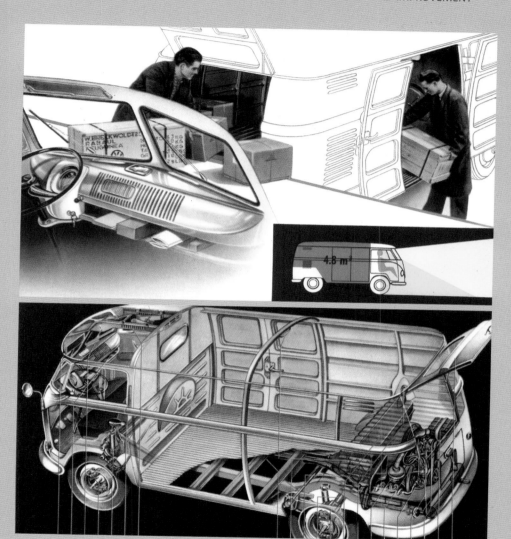

It was left to artists other than Reuters to portray the reality of the changes that occurred in March 1955. One image portrays both the new full-length metal dashboard and the rear-opening hatch, while the other, a cutaway diagram, highlights the space available above the engine and the spare-wheel housing behind the cab bench seat, not to mention part of the structure which made the vehicle appropriately rigid.

The Double Cab Pick-up proved popular from its earliest days. This line up of seven such vehicles was due to act as back-up vehicles for autobahn construction near Munich.

1961, September	Production started of a high roof version of the Delivery Van, the essence of which was extended body panels above the vehicle's waistline and taller side loading doors. The hatch at the rear remained the same size as that of the standard Delivery Van and a filler panel sat above this and the vehicle's roof. Similarly, at the front a curved panel sat above the split-panel windscreen and the cab doors.
1962, July	Single cab bench seat replaced by single adjustable seat for the driver and a bench for two passengers.
1963, January	42PS, 1500 engine optional for US market and, from March, passenger carrying models elsewhere. From August available on all models. 1200 still available, but deleted October 1965 due to lack of demand.
1963, August	Wider rear window in enlarged tailgate.
1963, December	15in wheels now universally replaced by 14in versions.
1964, August	Carburettor throttle governor fitted after it was discovered owners were exceeding the maximum recommended speed of 65mph by as much as 10mph.
1966, August	12V electrics standard, previously 6V.

Above: The High Roof Delivery Van being loaded with stock from a high street clothes store. Uses such as this were intended when the vehicle was added to the range.

A victim of the new larger hatch and rear window! The Micro Bus De Luxe, pictured here in a dealer's showroom, no longer carried its two curved upper rear quarter panel windows, due to lack of available room.

The front cover of a brochure presented at the time of the arrival of a larger hatch and increased rear window. Such was the Transporter's popularity that it was no longer considered necessary to portray the whole vehicle on the cover, or to clutter the image with a title!

The back cover of the same brochure showing the new, larger tailgate open, while the photographer also captures the increase in the size of the rear window to maximum effect.

The VW Camper

INTRODUCTION

Two rather obvious clues regarding the origins and nature of the VW Camper have been sprinkled across the pages of the preceding chapters. First, although Nordhoff did not decree that a Camper be included in his list of Transporter prototypes, his instruction that the Kombi was a key requirement sowed the seed for a vehicle that could act as a Delivery Van during the week while providing overnight accommodation, washing and cooking facilities for weekends away.

Second, the evermore-official endorsement of the creation of special use vehicles, largely developed by coach-building firms, and culminating in the issuing of SO numbers, opened the door for the karosserie Westfalia to partner Volkswagen in Germany and beyond. Here in the UK, a similar arrangement of partnership working between dealers and cabinet makers, or even builders, ensured that a right-hand-drive market, which Westfalia opted to ignore, was not starved of what would become an increasingly sought after commodity.

However, in case the point has been missed, Volkswagen did not manufacture its first Camper until the debut of the fifth-generation Transporter in the new millennium. The term VW Camper used previously was simply a convenience; an expediency that eventually came to encompass all models of Volkswagen as far as an element of the general public is concerned.

WESTFALIA

Although it would be wrong to assume that others did not create Campers in the early days of the Transporter, such are the fortunes of the Westfalia-Werke interwoven with Volkswagen, both at home and abroad, that to tell any story other than theirs would be unjust.

With origins dating back to 1844, Westfalia had a long history of producing carriages and trailers and, during World War 2, truck platforms for Ford chassis. Although the setback of a bomb-damaged and then levelled factory towards the war's end was significant, Westfalia had bounced back sufficiently by 1948 that it was in a position to exhibit both trailers and a small steel-plated caravan at the Hanover Fair. By 1950, the payroll stood at 300 with more expected to join very soon.

Ihr Landhaus auf Rädern!

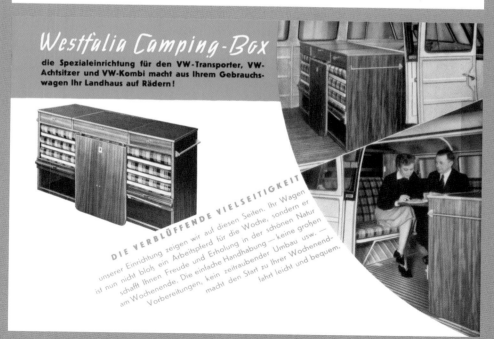

Westfalia Camping-Box

die Spezialeinrichtung für den VW-Transporter, VW-Achtsitzer und VW-Kombi macht aus Ihrem Gebrauchswagen Ihr Landhaus auf Rädern!

DIE VERBLÜFFENDE VIELSEITIGKEIT

unserer Einrichtung zeigen wir auf diesen Seiten. Ihr Wagen ist nun nicht bloß ein Arbeitspferd für die Woche, sondern er schafft Ihnen Freude und Erholung in der schönen Natur am Wochenende. Die einfache Handhabung — keine großen Vorbereitungen, kein zeitraubender Umbau usw. — macht den Start zu Ihrer Wochenendfahrt leicht und bequem.

The Knöbel brothers, then running the business, therefore took scant interest in an American officer serving in Germany who approached the company with a request for them to build a bespoke caravan style interior for a VW Kombi. However, such was the appeal of the finished product, that during 1952 Westfalia handcrafted a further 50 Camper conversions, having opened discussions with Wolfsburg regarding availability of vehicles.

Around the same time, the Knöbel brothers decided to offer what they called a Camping Box, a cabinet that could be dropped into a Kombi with little effort for the weekend and removed again on a Monday morning. The Camping Box was designed to fit against the front bulkhead in the load or passenger area. Its make-up consisted of three horizontal sections, the uppermost itself divided into three. The outer divisions contained drawers, while the centre section concealed a two-burner stove, accessed via a hinged lid. A rail positioned at one end of the unit held tea cloths or towels. The middle section, or shelf, was home to three boards, which formed the base for either seating or beds, and what Westfalia described as four part mattresses, which acted as cushions. A fold-down table, attached to the top of this centre section, doubled as a mattress board. The section closest to the floor was reserved for the storage of utensils. Ancillary fittings were also made available and included a vanity cupboard (which could be attached to the rear side-loading door) and a linen cupboard intended for positioning on the shelf above the engine, which being rather large completely obscured the vehicle's rear window.

From these humble origins, regular updates, particularly of the fully fitted out Campers, ensured ever-increasing sales, while from what had once simply been the 'Volkswagen Kombi with Westfalia De Luxe Camping Equipment', there emerged the beautifully crafted and officially designated SO23. This was replaced in the spring of 1961 by the SO34 and SO35 (the difference in number pertaining to white and grey laminate, or dark Swiss pear wood). These Campers were in turn replaced by the SO42 and SO44 in 1965, the two models on this occasion sporting very different layouts.

Shortly after the first-generation Transporter was replaced by the second, on 8 March 1968, Westfalia celebrated the occasion of the completion of the 30,000th conversion. In more recent times, demand had consistently outstripped supply, the biggest sufferers being US dealers, who had little option but to take drastic steps of their own.

THE VW CAMPER IN AMERICA

Once the relationship between Hanover and Westfalia had been established, Volkswagen's giant publicity machine swung into action. By the late 1950s, it was difficult to distinguish between a Camper produced by Westfalia and a model manufactured at Hanover alone. Initially described as 'The Volkswagen Camper with Westfalia De Luxe Equipment' and accompanied by Volkswagen's familiar logo, by the early 1960s the description had been simplified to 'The Volkswagen Camper'. However, lurking amongst

Two images from Westfalia's 1957 brochure, where the company presented its 'holiday home on wheels' with deluxe equipment but based on the VW Kombi.

Three images taken from Volkswagen's own brochure designed to promote the SO23. Creating an attractive cover for such a brochure could only be done effectively by depicting the VW and some of the trappings associated with camping. Fortunately, help was at hand for the serious purchaser, as convenient layout drawings for both day and nighttime graced the interior pages.

The cover of a brochure dedicated to the SO34 (pictured) and SO35. On this occasion, Volkswagen illustrated some of the vehicle's equipment through what are now delightful period images of a family on holiday in their camper — wash time and 'let's get cooking' being the two scenes depicted here.

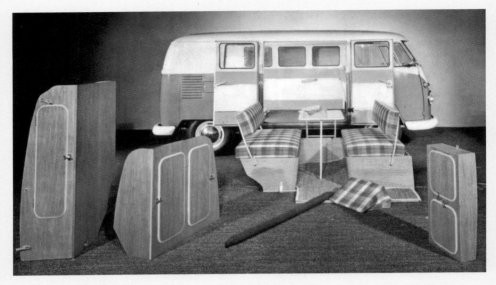

Although the Camping Box as such had been replaced, it was still possible to purchase a kit that could be easily removed and put back in again. This is the SO22 VW Camping Mosaik.

Moving forward a few years, these two images portray the SO42 and the SO44, although few but the most dedicated of enthusiasts would know which is which without additional imagery. Volkswagen duly supplied this with lifestyle pictures showing little more than a glimpse of the respective Campers' layouts. The SO42's owners are sat outside if that helps!

such publications were others such as *Getaway Car* and *Send this kit to Camp*, brochures produced by Volkswagen of America, to promote what was becoming known as the Campmobile and in later years by enthusiasts as the Westfakia!

Send this kit to camp

The exact nature of the reasons why VWoA or its dealers requested firms to build Campmobiles is a little unclear. In addition to the story already alluded to of demand outstripping supply, it may be that the workhorse elements of the Transporter range were slower sellers and that Volkswagen in Germany insisted America still took its quota to qualify for a supply of Campers. Whatever, the truth of the matter, either VWoA or the dealers commissioned the RV firm of Sportsmobile to create kits similar in layout and detail to Westfalia's, but including parts specific to Delivery Van installation. These were rushed to the ports, where, when a consignment of Delivery Vans arrived, work could start straight away.

Three images from the VW of America brochure entitled *Send this kit to Camp*, a publication designed to promote what some opted to call the Westfakia and one which launched the term 'Campmobile'. Despite the attractive imagery, the vehicle's origins as a Delivery Van are evident thanks to the style of the side windows.

By the end of 1963, Westfalia's ability to supply had improved somewhat, but possibly under the guidance of Volkswagen's advertising agency DDB, the use of the name Campmobile was encouraged and remained in use throughout the remaining years of first-generation production and over the lifespan of the second-generation model.

VOLKSWAGEN CAMPERS IN BRITAIN

While as the years passed by neither Germany nor the United States relied solely on the products emanating from Westfalia, or in the latter instance from VWoA/dealer copies, the story of the emergence of the VW Camper in Britain is altogether more diverse. While Peter Pitt (later of the popular Canterbury Pitt conversion fame) defied legislation, which would have made Camper ownership impractical and brought about a change in the law, other operations opened for business often to disappear within a few years.

Notable amongst these were Volkswagen distributors Moortown Motors, whose range of vehicles soon carried Autohome branding. Moortown employed a local cabinet making firm to produce the interiors, while the distributor was responsible for sales and marketing. Remarkably sophisticated for their time, although a budget version Campahome was offered towards the end of production, success was not forthcoming and, as a result, the final output was manufactured in October 1963.

As the 1960s unfolded the well-known firm of Martin Walter, with its redoubtable Dormobile brand, was one of the first to add Volkswagen Campers to its already extensive range of Fords, Morris, Commers and even Land Rovers, while mid-decade Essex-based Danbury produced its first Multicar, using a second-hand Kombi for the job. Both these players would survive for a considerable time, Dormobile finally exiting the scene towards the end of second-generation Transporter production in 1979, and Danbury during the run of the third-generation models in the latter half of the 1980s. By way of a footnote, Canterbury Pitt's Volkswagen Open Plan Moto-Caravan ceased production on the death of Peter Pitt in February 1969, a consequence of a complicated licensing arrangement, which expired on his demise.

Pioneer Peter Pitt's Moto-Caravan (seen here in 1959/60) was quite sophisticated for its time. Later examples (after Peter Pitt had 'merged' with Canterbury) remain sought after to this day.

MOORTOWN MOTORS LTD. REGENT STREET, LEEDS 2.   Tel. 31894

How times change. While the interior of the Moortown Motors' Autohome was well appointed for both dining and sleeping, as this photograph illustrates, dress-sense was completely different to that of more recent times. Ties, suits, pearl necklace and earrings were soon to be a thing of the past as far as leisure activities went.

Leeds-based Moortown Motors appeared to be making headway from their debut in 1958 with new models virtually every year. However, like many of the other early players, success did not last.

Although conversion companies came and went, one operation appeared set to stay the course. In typical British fashion, the story of Devon's emergence on the Camper scene is a fascinating one. Sidmouth-based builder Jack White bought a Transporter when the family, now numbering three children, outgrew his Beetle. Long journeys to visit his wife Anna's relations in Germany suggested more comfort was required and, in the depths of the winter of 1955, Jack planned an upgrade for the Delivery Van. During the course of February 1956 and with the help of his designer and builder of kitchen units, Pat White, a bespoke interior was created for the vehicle.

Working from Jack's garden shed, the two men built four seats and finished a laminated worktop. These were designed to convert into a double divan in the centre section of the vehicle, while a single bed was added over the engine bay and another crossways utilising the front bench seat. A little later a fifth bed was added, which, with considerable ingenuity of design, was suspended above the bottom end of the double divan. The Transporter's makeover extended to the installation of a cooker, basic washing facilities, an Osokool storage cabinet, fitted cupboards and both Calor gas and 6V dc electricity. Luxury Dunlopillo cushions and neatly fitting curtains completed the arrangements.

To avoid Purchase Tax, the completed vehicle had to be approved by ministry officials as living accommodation. Jack took his Camper to the relatively nearby Volkswagen main dealer, Lisburne Garage Ltd in Torquay, where it awaited inspection. Although parked out of the way, the vehicle attracted so much attention that it was suggested it should be put on display in the main showroom. Before long, the garage had a number of serious enquiries, which in turn led to serious consideration being given to commercial production.

The Devon CARAVETTE

the ideal holiday home

MOTORISED CARAVAN ON
THE VOLKSWAGEN MICRO-BUS

The cover of Devon's 1960 brochure was unusual in that the family group posing by the Caravette consisted of founder Jack White, his wife Anna and their three children.

With the ministry inspection completed, Jack determined how he wished to proceed and in July 1956, using the brand name Devon, in honour of his home county and christening the layout design as the Caravette, J P White (Sidmouth) Ltd, makers of Camper vans was in business. By the end of the year, 56 vehicles had been sold through Lisburne Garages, who were responsible for sales, distribution and after-sales service on Jack's behalf.

A period of intense activity followed demanding more than one relocation as production outgrew the premises. Finally, Jack constructed a purpose-built works, which, when it opened on 20 May 1960, was expected to be able to cope with annual production of around 1,000 vehicles per year and house a workforce of 75.

Sadly, Jack White collapsed and died in the arms of Pat Mitchell in November 1963 at the tragically early age of 51. A year later, the business was sold to the Renwick, Wilton & Dobson Group, which decided to retain both the Devon name and the exceptionally high standards of craftsmanship associated with it.

For many a year Devon conversions were crafted in hand-polished solid oak, with models manufactured before 1961 featuring a gorgeous curved side cabinet to the rear of the side loading doors. Although this unit disappeared from the specification, it only did so as Devon strove to improve their layouts on a year-by-year basis. Their 1962 range also included a budget model named the Devonette, which would later give way to a second economy model, the Torvette that was launched in October 1965. Caravette conversions tended to be on either the Micro Bus or even the Micro Bus De Luxe, while the budget models could be specified on the Kombi as well as the Micro Bus.

The Devon models had become Britain's best selling conversions and rightly so! The promise of a second generation Transporter heralded further and greater glories.

Part of the cover of Devon's 1967 brochure that carried the following message on its cover: 'A hardworking Devon motor caravan earns its holiday in the sun.' Although Devon never produced a Westfalia style removable kit, it sold the message hard that its Caravette and other models could be used all week.

Eventually black and white gave way to colour on the cover of the Devon brochures. This example dates from the autumn of 1962 and introduces the company's 1963 range. As usual, the interior pages gave details of the layout and models, the cover acting as little more than an eye-catcher in a dealer's showroom.

By the end of first generation Transporter production, Devon was offering both a deluxe (Caravette) and more basic (Torvette) model, plus conversions based on full bulkhead and walkthrough models.

The Devon 'Caravette' Spaceway Model

This latest addition to the Devon range offers the famous quality and finish of the Standard model plus the 'Spaceway' access connecting the driving-cab with the interior. Another unique feature is the 'swing-out' unit comprising two-burner cooker with grill, 7-gallon water tank with hand-pump, storage for bottled gas etc. and useful drawer. The 'Spaceway' model also differs from the Standard as follows: □ one large table adaptable for use outdoors; □ full-length double bed (alternatively a full-length single bed) and a single bunk for one child, giving excellent sleeping accommodation; □ optional extra hammock bunks available for one or two children; □ large-capacity 'Easikool' evaporation food cooler with built-in shelf.

The Devon 'Torvette' Standard Model

This conversion, on the Volkswagen Kombi, offers exceptional quality in the lower price range. Its remarkable adaptability provides a choice of a single bed, a double bed, or two single beds for adults, plus sleeping accommodation for two children. The carriage of bulky items, if required, is catered for by the ingenious conversion which makes a large floor space readily available. All the necessary equipment is included in this model, plus many of the famous Devon refinements, thus providing a fully comprehensive motor caravan for every purpose.

The VW Fleetline

As the saying goes, all good things come to an end! For the first-generation Transporter, nurtured and subject to continual improvement by its mentor Heinz Nordhoff for over 13 years, its death knell was sounded in 1964. As with his other great love, the much older, indeed pre-war, Volkswagen Beetle, Nordhoff was occasionally presented with designs and plans for his protégé's replacement. Nor were such offerings the machinations of would-be revolutionaries – ever the realist, they came before the Director General because he had asked for them.

The first of the would-be first generation Transporter replacements was presented to the Director General as early as 1960. Code named EA 114 this vehicle had reached the prototype stage before Nordhoff dismissed it. Three years later with his fellow directors' support, Nordhoff charged Gustav Mayer, head of the development department for commercial vehicles, to work towards a new version of the Transporter. Towards the end of 1964 the project was given full go ahead for series production. Those who would later say Nordhoff was incapable of changing one model for another, carefully overlook the fact that the conception and birth of the second-generation Transporter both took place in his lifetime. Nordhoff axed the first generation Transporter because he realised its future success was no longer assured. As if to prove his point, sales dropped from their all-time peak in 1964 the following year and never recovered. Conversely, in 1965, Beetle production reached a new high with over a million cars produced for the first time in a single 12-month period.

Although the first generation Transporter disappeared from German and most other showrooms across the world in the summer of 1967, its story was not yet over. The increasingly maverick Brazilian operation decided it was not changing and its São Paulo factory continued to produce the first-generation Transporter until the autumn of 1975. (Even then, some would argue that the model that replaced it was little more than a first-generation Transporter with a second-generation front welded on to it!)

Volkswagen do Brasil SA had come into being in March 1953 and immediately started producing Beetles and Transporters from CKD kits supplied by Wolfsburg. Before turning to full manufacture in September 1957, Brazil assembled a hardly earth-shattering 552 Kombis, as all models were known in the country. However, having turned to manufacturing, numbers rose as the chart below indicates. Even more significantly, at the point Hanover stopped producing the first-generation Transporter, Brazilian output increased dramatically! Additionally, from 1971 Brazil started to supply CKD kits to other countries, primarily but not exclusively, in Latin America.

Running out of ideas? This brochure cover, which dates from 1965, depicts some of the many Transporter variations offered. Note particularly: row one — Double Cab (before spotting many more); row two — Pick-up with wide metal platform; row three — High Roof Delivery Van; row four — Fire Wagon.

Amidst the pages of a South African market brochure dedicated to second-generation Transporters, browsers would have come across this image of the VW Fleetline, a Brazilian Transporter assembled in South Africa — the year was 1975.

PRODUCTION OF THE KOMBI

Year	Production	Additional CKD Kits
1957	371	
1958	4,819	
1959	8,383	
1960	11,299	
1961	16,315	
1962	14,563	
1963	14,428	
1964	12,378	
1965	13,114	
1966	15,138	
1967	21,172	
1968	26,883	
1969	28,253	
1970	30,205	
1971	28,316	444
1972	34,898	720
1973	44,083	1,536
1974	48,803	3,540
1975*	53,335	6,060

*includes months after October when production had changed to new model

Brazil's Kombi varied a little in specification from Hanover's first-generation Transporter. For example, the 1500 engine was only fitted from 1967, while passenger-carrying· vehicles persisted with the wraparound upper rear quarter panel windows associated with the German Micro Bus De Luxe until a larger tailgate was introduced in August 1963.

The non-South American recipient of Brazilian CKD kits was South Africa. Alongside the new second-generation Transporter, the Uitenhage factory assembled first-generation models in three guises. In addition to a Delivery Van and Pick-up, there was also a multi windowed Kombi. All were sold as budget models under the brand heading of 'Fleetline'. The numbers assembled were low, one source suggesting a total of no more than 789 vehicles before the end of production in 1975, but many appear to have survived, giving today's enthusiasts the opportunity to purchase a modern first generation Transporter!

EPILOGUE

From a speech made by Heinz Nordhoff on the occasion of the production of the millionth Transporter:

'... The VW Transporter was something completely new and original. No one foresaw that it would become the forerunner of a completely new type of vehicle, representing as it did an often near-slavishly imitated genre of utility vehicles.

'When one mentions vehicles, one usually thinks of cars, but production of these small Transporters of up to 1,000kg represents a considerable industry in its own right, probably the newest in the entire automobile industry, both in Europe and the USA ...

'Sixty three per cent of all Transporters are exported and their market share in this newly created sector is very high in many countries. It is 40.5 per cent in this country, but in Belgium, Holland, Austria, Sweden and Switzerland it is 50 per cent and more ... The VW Transporter is not only the original, but also remains the leader in its class.

'In the relatively short time of less than 15 years, something new has been created by the VW Factory, of which it can be said "Often copied, never equalled".'

Other pages of the same brochure offered further variations on the Fleetline theme — here's the double page spread dedicated to the first-generation Brazilian Pick-up.

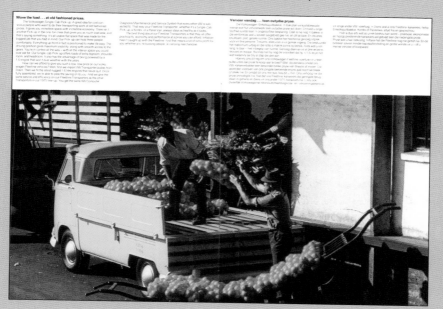

Index